Math Word Problems

Grade 3

Written by
Annette Taulbee

Illustrated by
Lynn Conklin Power

Notice! Copies of worksheets may be reproduced by the classroom teacher for classroom use only, not for commercial resale. No part of this publication may be reproduced for storage in a retrieval system, or transmitted in any form or by any means—electronic, mechanical, recording, etc.—without the prior written permission of the publisher. Reproduction of these materials for an entire school or school system is strictly prohibited.

Copyright © 1988 Frank Schaffer Publications, Inc.
All rights reserved— Printed in the U.S.A.
Published by **Frank Schaffer Publications, Inc.**
1028 Via Mirabel
Palos Verdes Estates, California 90274

Name _____ Skill: Multiplication

Write the number sentence and label your answer.

1. There are 3 spiders. If each spider has 8 legs, how many legs are there in all?

$3 \times 8 = 24$

There are __24__ __legs__.

2. There are 3 dogs. If each dog has 4 legs, how many legs are there altogether?

_____

There are _____ _____.

3. If there are 5 ants, each with 6 legs, how many legs are there?

_____

There are _____ _____.

4. Ducks have 2 feet. If there are 7 ducks, how many feet are there in all?

_____

There are _____ _____.

5. An octopus has 8 arms. How many arms would there be if you had 4 octopuses?

_____

There would be _____ _____.

6. Snakes don't have legs. If there are 3 snakes, how many legs are there?

_____

There are _____ _____.

7. If one cricket has 6 legs, how many legs would 3 crickets have altogether?

_____

There would be _____ _____.

© Frank Schaffer Publications, Inc. FS-8552 Math Word Problems

Name _____ Skill: Multiplication

Write the number sentence and label your answer.

1. A football team gets 6 points for a touchdown. How many points would 8 touchdowns be?

_____

It would be _____ _____ .

2. Baseball games have 9 innings. If each inning has 3 outs, how many outs are there in a game?

_____

There are _____ _____ .

3. If you swam 8 laps a day for 7 days, how many laps would you swim altogether?

_____

That would be _____ _____ .

4. There are 4 people at each ping pong table. If there are 9 tables, how many people are playing?

_____

There are _____ _____ .

5. Dad jogs 6 miles a day. How many miles would he jog in 4 days?

_____

Dad would jog _____ _____ .

6. There are 5 players on a basketball team. If they each score 9 points, what would the team's score be?

_____

The score would be _____ .

7. In tennis, you play 6 games each set. If you play 7 sets, how many games would you play?

_____

You would play _____ _____ .

Name _____   Skill: Division

Write the number sentence and label your answer.

1. John has 20¢. If donuts are 5¢ each, how many can he buy?

$$20 \div 5 = 4$$

John can buy __4__ __donuts__.

2. The baker made 72 brownies. He put 9 rows of them on a tray. How many were in each row?

_____

Each row had ____ _____.

3. Sue has 48¢. If cookies are 8¢ each, how many can she buy?

_____

Sue can buy ____ _____.

4. If Jill buys 24 cream puffs and divides them equally into 4 bags, how many will be in each bag?

_____

Each bag will have ____ _____.

5. The baker made 12 cake layers. He needs to make three cakes. How many layers will each cake have?

_____

Each will have ____ _____.

6. Bill has 49¢. If donut holes are 7¢ each, how many can he buy?

_____

Bill can buy ____ _____.

7. The baker put 56 hot rolls in 8 rows. How many were in each row?

_____

Each row had ____ _____.

Name _____ Skill: Division

Write the number sentence and label your answer.

1. There were 16 boy scouts. They slept in 8 tents. How many boys were in each tent?

_____

Each tent had _____ _____ .

2. If 32 girl scouts stayed in 4 cabins, how many were in each cabin?

_____

Each cabin had _____ _____ .

3. There are 6 large tables in the mess hall. If there are 54 scouts, how many will sit at each table?

_____

_____ _____ at each table.

4. 63 girl scouts wanted to go boating. If there are 9 boats, how many girls will be in each boat?

_____

_____ _____ in each boat.

5. Seven groups of boy scouts went hiking. There were 42 boys. How many were in each group?

_____

Each group had _____ _____ .

6. If 25 scouts make 5 campfires, how many will sit at each campfire?

_____

_____ _____ at each campfire.

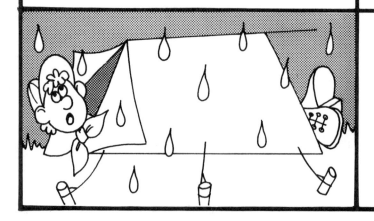

7. The arts and crafts leader can take groups of 9 scouts at a time. If 27 scouts sign up, how many groups would come?

_____

There would be _____ _____ .

Name _____ Skill: Multiplication or division

Write the number sentence and label your answer.

1. Jack planted 3 rows of apple trees, with 6 trees in each row. How many apple trees does he have?

_____

Jack has ____ _____ .

2. Mary divided 10 peaches into 2 baskets. How many peaches are in each basket?

_____

____ _____ in each basket.

3. If Sam planted 21 orange trees in 7 rows, how many trees are in each row?

_____

Each row had ____ _____ .

4. Sue had 9 baskets. If she put 7 apples in each one, how many apples did she pick?

_____

Sue picked ____ _____ .

5. Bill had 7 peach trees. He picked 4 good peaches from each tree. How many peaches did he have in all?

_____

Bill had ____ _____ .

6. John divided 30 oranges equally between 6 friends. How many oranges did each friend get?

_____

Each friend got ____ _____ .

7. Betty has 63 apples to put into 7 baskets. How many apples did she put in each basket?

_____

Each basket had ____ _____ .

Name _____ Skill: Multiplication or division

Write the number sentence and label your answer.

1. Jim has 36¢. If candy bars are 9¢ each, how many can he buy?

_____

Jim can buy _____ _____ .

2. Sue has 7¢. Gumdrops are 1¢ each. How many can she buy?

_____

Sue can buy _____ _____ .

3. Bill bought 5 jawbreakers for 4¢ each. How much did he spend?

_____

Bill spent _____ _____ .

4. Candy mints are 3¢ each. If Nancy has 6¢, how many can she buy?

_____

Nancy can buy _____ _____ .

5. Jack bought 2 packages of gum. There are 8 pieces of gum in each package. How many pieces did he have?

_____

Jack had _____ _____ .

6. Bob wants to buy 6 candy bars. They are 9¢ each. How much money does he need?

_____

Bill needs _____ _____ .

7. The candy man has 8 jars of candy sticks. Each jar has 5 sticks of candy. How many are there in all?

_____

There are _____ _____ .

Name _____   Skill: Multiplication or division

Write the number sentence and label your answer.

1. How much more will Bill spend than Sue if:

   a. Bill buys 5 oranges for 8¢ each? _____

   b. Sue buys 4 apples for 9¢ each? _____

   Bill will spend _____ more than Sue.

2. Which is the better buy? (Circle a or b)

   a. Sue's 4 small cans of juice at 7¢ each. _____

   b. Bill's 3 cans of juice at 9¢ each. _____

   How much better? _____

3. What is the cost per pear if:

   a. Bill spends 54¢ for 9 pears. _____

   b. Sue spends 28¢ for 4 pears. _____

   _____ spends _____ more per pear.

4. How much less will be spent per lemon if:

   a. Sue spends 64¢ on 8 lemons. _____

   b. Bill spends 25¢ on 5 lemons. _____

   _____ will spend _____ less per lemon.

5. How much difference in total price is there if:

   a. Bill buys 6 carrots for 7¢ each. _____

   b. Sue buys 5 potatoes for 9¢ each. _____

   There is _____ difference in price.

© Frank Schaffer Publications, Inc.   FS-8552 Math Word Problems

Name _____ Skill: 3 digit addition

Write the number sentence and label your answer.

Distances:
- Dry Gulch to Ghost Town: 62 miles
- Dry Gulch to Strike it Rich: 124 miles
- Dry Gulch to Miner's City: 202 miles
- Dry Gulch to (diagonal): 313 miles
- Ghost Town to Miner's City: 75 miles
- Strike it Rich to Miner's City: 151 miles

---

1. The Browns drove from Dry Gulch to Strike it Rich. Then they drove to Miners City the next day. How many miles did they drive?

313 + 151 = 464

They drove **464 miles**.

---

2. From Miners City, the Browns headed for Ghost Town. Then they went back to Strike it Rich. How many miles did they travel?

_____

They traveled _____ _____.

---

3. The Williams family drove from Ghost Town to Dry Gulch, and then on to Miners City. How many miles did they travel?

_____

They traveled _____ _____.

---

4. From Miners City, the Williamses drove to Strike it Rich, and then to Ghost Town. How far did they go?

_____

They went _____ _____.

---

5. How many miles is it from Strike it Rich to Ghost Town, to Dry Gulch, and then to Miners City?

_____

It is _____ _____.

---

6. How long a trip would it be from Strike it Rich to Dry Gulch, to Miners City, and back to Strike it Rich?

_____

It would be _____ _____.

Name _____ Skill: 3 digit subtraction

Write the number sentence and label your answer.

1. John saved $5.95. If he buys the Rainbow Game, how much will he have left?

$5.95 - 3.80 = 2.15$

John will have ___$2.15___ left.

2. If Ann gives the sales person $5.49 to pay for the race car, how much money will she get back?

_____

Ann will get _____ back.

3. Tammy wanted to buy the walkie-talkies. Her dad gave her $9.87. How much money will be left over?

_____

There will be _____ left.

4. If Jim had $9.68 and he bought a Frisbee, how much money would he have then?

_____

Jim would have _____ .

5. Mark has $5.69. If he buys a race car, how much will he still have?

_____

Mark will still have _____ .

6. If Jan had $9.85 and bought the Rainbow Game, how much would she have left?

_____

Jan would have _____ left.

Name _____ Skill: 3 digit addition and subtraction

Write the number sentence and label your answer.

1. How many people live in the towns of Rice and Upton combined?

_____

There are _____ _____ .

2. If 250 people move out of Newark, how many will be left?

_____

_____ _____ will be left.

3. If all the people in Blair and Upton get together, how many would there be?

_____

There would be _____ _____ .

4. If 131 of the people in Rice go on vacation, how many will still be there?

_____

_____ _____ will still be there.

5. If 102 people in Upton move, how many will be left?

_____

_____ _____ will be left.

6. How many people live in Upton, Blair, and Rice combined?

_____

There are _____ _____ .

Name _____ Skills: Comparisons (3 digit)

Write the number sentence and label your answer.

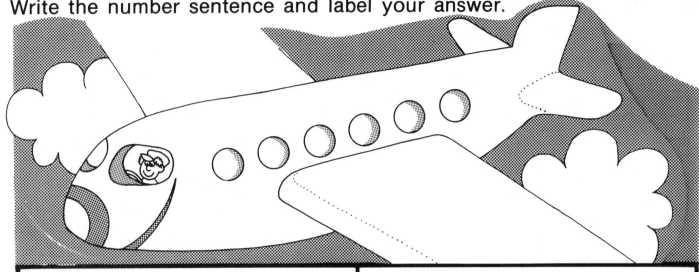

1. Jan's airplane tickets cost her $463. Sue's were $586. How much more did Sue spend than Jan?

   $586 - 463 = 123$

   Sue spent __$123__ more.

2. Bill flew 375 miles on Monday and 130 miles on Tuesday. How much farther did he fly on Monday than Tuesday?

   _____

   Bill flew ____ ____ more.

3. Captain Barns flies at 678 mph. Captain Frost flies at 361 mph. How much faster is Captain Barns' plane?

   _____

   His plane is ____ ____ faster.

4. Jack flew 264 miles. Mary flew 784. How many more miles did Mary fly?

   _____

   Mary flew ____ ____ more.

5. A trip to Hawaii is $898. A skiing trip is $465. How much more is the trip to Hawaii?

   _____

   Hawaii is ____ ____ more.

6. A 747 airplane can seat 366 people. A DC-10 can carry 260 people. How many more people can fly on a 747?

   _____

   ____ ____ more.

Name _____ Skill: Review of 3 digits

Write the number sentence and label your answer.

16oz. Steak..... $4.58
Italian Spaghetti... $2.24
Salad..... $1.43
Hamburger $1.32
Milk .31
Coffee .40

"May I take your order, please?"

1. Dad had steak and a cup of coffee for dinner. How much was his bill?

_____

Dad's bill was _____ .

2. Mom had the spaghetti and a salad. How much did she spend?

_____

Mom spent _____ .

3. John had $3.68. If he bought a hamburger, how much did he have left?

_____

John had _____ left.

4. Sue had a salad and a glass of milk. If she left a 25¢ tip, how much did she spend?

_____

Sue spent _____ .

5. How much more did Dad's dinner cost than Mom's?

_____

It cost _____ more.

6. How much difference in price is there between the steak dinner and the spaghetti dinner?

_____

A difference of _____ .

Name _____   Skill: 3 digit addition-regrouping

Write the number sentence and label your answer.

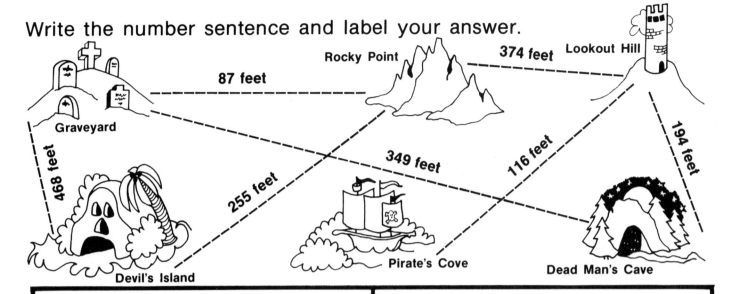

1. Sue and Bill started at Dead Man's Cave, walked to the Graveyard and then to Devil's Island. How far did they go?

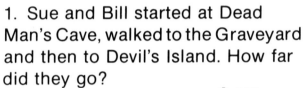

They walked 817 feet .

2. John and Jim went from Dead Man's Cave to Lookout Hill and then to Pirate's Cove. How far had they gone?

_____

They went _____ _____ .

3. If Nancy and Ann walked from Devil's Island to Rocky Point and then to the Graveyard, how long was their trip?

_____

Their trip was _____ _____ long.

4. Tom started at Devil's Island and headed for the Graveyard. From there, he walked to Rocky Point. How many feet did he walk?

_____

Tom walked _____ _____ .

5. How far is it from Lookout Hill to Rocky Point and back again?

_____

It is _____ _____ .

6. How many feet is it from Rocky Point to the Graveyard if you go by Devil's Island?

_____

It is _____ _____ .

© Frank Schaffer Publications, Inc.   FS-8552 Math Word Problems

Name _____ Skill: 3 digit subtraction-regrouping

Write the number sentence and label your answer.

1. Mike had $6.91. He bought a shirt. How much did he have left?

   $6.91 - 5.87 = 1.04$

   Mike had ___$1.04___ left.

2. Linda bought the skirt. If she gave the sales person $9.70, how much change did she get?

   _____

   Linda got _____ change.

3. Jim's dad gave him $9.25. If Jim bought the pants, how much money will his dad get back?

   _____

   He will get back _____.

4. Mark bought the boy's hat. If he started with $4.50, how much does he have left?

   _____

   Mark had _____ left.

5. Ann's mother had $9.79. If Ann buys the blouse, how much change will her mother get?

   _____

   She will get _____ change.

6. If Nancy buys the girl's hat, how much change will she get from a $5.00 bill?

   _____

   Nancy will get _____ change.

Name _____   Skill: 3 digit addition and subtraction-regrouping

Write the number sentence and label your answer.

1. Andy made $2.75 on Saturday and $3.25 on Sunday. How much did he make altogether?

_____

Andy made _____ .

2. Mike saved $7.82. He spent $4.95 on a gift for his brother. How much did he have left?

_____

Mike had _____ left.

3. Susan spent $6.84 on a gift for her mother and $1.77 to get it wrapped. How much did she spend in all?

_____

Susan spent _____ in all.

4. Bob bought 3 toys. One was $1.49, another was $2.37, and the third was $.38. What was the total?

_____

Total cost was _____ .

5. Mary had $9.00. She paid $2.64 for her sister's birthday gift. How much did she have then?

_____

Mary had _____ then.

6. Jack saved $3.49. On the way to the store, he lost $.27. How much can he spend now?

_____

Jack can spend _____ now.

7. If Linda started with $8.70 and spent $1.57 on her baby sister, how much money does she have left?

_____

Linda has _____ left.

Name _____ Skill: 3 digit-regrouping—comparisons

Write the number sentence and label your answer.

1. How much more does the hippo weigh than the lion?

   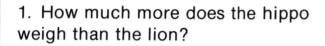

   _ 657 _ _ pounds _ more.

hippo	924 pounds
bear	826 pounds
gorilla	342 pounds
zebra	325 pounds
lion	267 pounds
seal	129 pounds
wolf	102 pounds

2. How much less does the wolf weigh than the gorilla?

   _____

   _____ _____ less.

3. How much difference in weight is there between the zebra and the seal?

   _____

   The difference is _____ _____ .

4. How much difference in weight is there between the bear and the gorilla?

   _____

   The difference is _____ _____ .

5. How much less does the bear weigh than the hippo?

   _____

   _____ _____ less.

6. How much more does the gorilla weigh than the lion?

   _____

   _____ _____ more.

Name _____   Skill: Review-3 digits

Write the number sentence and label your answer.

Jim's Restaurant

Bacon & Eggs............$2.65
Ham & Eggs..............$3.25
Steak & Eggs............$4.80
Fresh Fruit................$1.39
Hash Brown Potatoes......90
Pancakes..................$1.75

Muffin..................30
Toast...................35
Cereal.................60
Milk....................45
Juice...................75
Coffee.................50

1. Mark had ham and eggs and a glass of milk for breakfast. How much did he spend?

   _____

   Mark spent _____ .

2. John wants bacon and eggs. How much less will his breakfast cost than Mark's?

   _____

   It will cost _____ less.

3. Ann had steak and eggs, toast and a glass of juice. How much was her bill?

   _____

   Ann's bill was _____ .

4. How much more do the ham and eggs cost than the pancakes?

   _____

   They cost _____ more.

5. Nancy had a muffin with her bacon and eggs. How much was her breakfast?

   _____

   Her breakfast was _____ .

6. How much difference in price is there between the pancakes and the fresh fruit?

   _____

   The difference is _____ .

Name _____ Skill: Review

Write the number sentence and label your answer.

1. Jim had $2.68 to spend. Linda had $3.45. How much did they have between them?

_____

They had _____ .

2. Jim bought 6 ride tickets. They were 10¢ each. How much was the total cost?

_____

Total cost was _____ .

3. Ann spent $2.60 at a booth. Tom spent $1.75. How much more did Ann spend?

_____

Ann spent _____ more.

4. Linda spent 40¢ on tickets. If they were 5¢ each, how many tickets did she get?

_____

Linda got _____ .

5. Mark spent $5.32 at the park. Mary spent $3.69. How much less did Mary spend than Mark?

_____

Mary spent _____ less.

6. John spent $2.65 on tickets, $1.98 on food, and $3.49 on a toy dog. How much did he spend in all?

_____

John spent _____ in all.

Name _____ Skill: Pre/Post Test

Write the number sentence and label your answer.

1. Mr Adams spent $4.69 for a saw, $2.88 for a hammer, and $1.63 for some nails. How much did he spend in all?

_____

He spent _____ in all.

2. Mr. Jones spent $6.42 on tools. Mr. Robbins spent $4.87 on tools. How much more did Mr. Jones spend than Mr. Robbins?

_____

He spent _____ more.

3. Tom bought 7 packages of hooks. If there were 8 hooks in each package, how many hooks did he buy?

_____

Tom bought ____ _____ .

4. Mrs. Adams looked at a step stool for $3.98 and one for $5.66. How much difference in cost was there?

_____

The difference was _____ .

5. Diane bought 225 large nails and 110 small ones. How many nails did she buy altogether?

_____

Diane bought ____ _____ .

6. Mrs. Robbins bought a package of screws for 48¢. If screws are 6¢ each, how many screws were the package?

_____

There were ____ _____ .

7. The hardware store had 582 gallons of paint in stock. During their sale, 189 gallons were sold. How many are left?

_____

There are ____ _____ left.

Yahoo!

I, _____,

have successfully completed a unit on WORD PROBLEMS using multiplication, division and 3 digit addition and subtraction. Just give me paper, a pencil and a problem. I'll solve it! No problem!

_____
Teacher

Answers

Page One
1. 3x8=24; 24 legs
2. 3x4=12; 12 legs
3. 5x6=30; 30 legs
4. 2x7=14; 14 legs
5. 8x4=32; 32 arms
6. 0x3=0; 0 legs
7. 6x3=18; 18 legs

Page Two
1. 6x8=48; 48 points
2. 9x3=27; 27 outs
3. 8x7=56; 56 laps
4. 4x9=36; 36 people
5. 6x4=24; 24 miles
6. 5x9=45; 45 points
7. 6x7=42; 42 games

Page Three
1. 20÷5=4; 4 donuts
2. 72÷9=8; 8 brownies
3. 48÷8=6; 6 cookies
4. 24÷4=6; 6 cream puffs
5. 12÷3=4; 4 layers
6. 49÷7=7; 7 donut holes
7. 56÷8=7; 7 hot rolls

Page Four
1. 16÷8=2; 2 boy scouts
2. 32÷4=8; 8 girl scouts
3. 54÷6=9; 9 scouts
4. 63÷9=7; 7 girl scouts
5. 42÷7=6; 6 boy scouts
6. 25÷5=5; 5 scouts
7. 27÷9=3; 3 groups

Page Five
1. 3x6=18; 18 apple trees
2. 10÷2=5; 5 peaches
3. 21÷7=3; 3 trees
4. 9x7=63; 63 apples
5. 7x4=28; 28 peaches
6. 30÷6=5; 5 oranges
7. 63÷7=9; 9 apples

Page Six
1. 36÷9=4; 4 candy bars
2. 7÷1=7; 7 gumdrops
3. 5x4=20; 20¢
4. 6÷3=2; 2 mints
5. 2x8=16; 16 pieces
6. 6x9=54; 54¢
7. 8x5=40; 40 candy sticks

Page Seven
1. 5x8=40; 4x9=36; 4¢
2. 4x7=28; 3x9=27; 1¢
3. 54÷9=6; 28÷4=7; Sue, 1¢
4. 64÷8=8; 25÷5=5; Bill, 3¢
5. 6x7=42; 5x9=45; 3¢

Page Eight
1. 313+151=464; 464 miles
2. 75+202=277; 277 miles
3. 62+124=186; 186 miles
4. 151+202=353; 353 miles
5. 202+62+124=388; 388 miles
6. 313+124+151=588; 588 miles

Page Nine
1. 5.95-3.80=2.15; $2.15
2. 5.49-1.49=4.00; $4.00
3. 9.87-6.66=3.21; $3.21
4. 9.68-2.53=7.15; $7.15
5. 5.69-1.49=4.20; $4.20
6. 9.85-3.80=6.05; $6.05

Page Ten
1. 472+215=687; 687 people
2. 987-250=737; 737 people
3. 300+215=515; 515 people
4. 472-131=341; 341 people
5. 215-102=113; 113 people
6. 215+300+472=987; 987 people

Page Eleven
1. 586-463=123; $123
2. 375-130=245; 245 miles
3. 678-361=317; 317 mph
4. 784-264=520; 520 miles
5. 898-465=433; $433
6. 366-260=106; 106 people

Page Twelve
1. 4.58+.40=4.98
2. 2.24+1.43=3.67; $3.67
3. 3.68-1.32=2.36; $2.36
4. 1.43+.31+.25=1.99; $1.99
5. 4.98-3.67=1.31; $1.31
6. 4.58-2.24=2.34; $2.34

Page Thirteen
1. 349+468=817; 817 feet
2. 194+116=310; 310 feet
3. 255+87=342; 342 feet
4. 468+87=555; 555 feet
5. 374+374=748; 748 feet
6. 255+468=723; 723 feet

Page Fourteen
1. 6.91-5.87=1.04; $1.04
2. 9.70-7.42=2.28; $2.28
3. 9.25-6.73=2.52; $2.52
4. 4.50-3.69=.81; $.81
5. 9.79-8.69=1.10; $1.10
6. 5.00-2.88=2.12; $2.12

Page Fifteen
1. 2.75+3.25=6.00; $6.00
2. 7.82-4.95=2.87; $2.87
3. 6.84+1.77=8.61; $8.61
4. 1.49+2.37+.38=4.24; $4.24
5. 9.00-2.64=6.36
6. 3.49-.27=3.22; $3.22
7. 8.70-1.57=7.13; $7.13

Page Sixteen
1. 924-267=657; 657 pounds
2. 342-102=240; 240 pounds
3. 325-129=196; 196 pounds
4. 826-342=484; 484 pounds
5. 924-826=98; 98 pounds
6. 342-267=75; 75 pounds

Page Seventeen
1. 3.25+.45=3.70; $3.70
2. 3.70-2.65=1.05; $1.05
3. 4.80+.35+.75=5.90; $5.90
4. 3.25-1.75=1.50; $1.50
5. .30+2.65=2.95; $2.95
6. 1.75-1.39=.36; $.36

Page Eighteen
1. 2.68+3.45=6.13; $6.13
2. 6x.10=.60; $.60
3. 2.60-1.75=.85; $.85
4. 40-5=8; 8 tickets
5. 5.32-3.69=1.63; $1.63
6. 2.65+1.98+3.49=8.12; $8.12

Page Nineteen
1. 4.69+2.88+1.63=9.20; $9.20
2. 6.42-4.87=1.55; $1.55
3. 7x8=56; 56 hooks
4. 5.66-3.98=1.68; $1.68
5. 225+110=335; 335 nails
6. 48÷6=8; 8 screws
7. 582-189=393; 393 gallons